The LIBRARY of LANDFORMS™

CAVES

Isaac Nadeau

The Rosen Publishing Group's
PowerKids Press™
New York

To Judy, Les, Gary and Barb

Published in 2006 by The Rosen Publishing Group, Inc.
29 East 21st Street, New York, NY 10010

First Edition

Editor: Rachel O'Connor
Book Design: Elana Davidian

Photo Credits: Cover, title page © Demetrio Carrasco/Getty Images; p. 4 (top) © Malie Rich-Griffith/infocusphotos.com; p. 4 (bottom) © Tom Bean/Corbis; pp. 7, 19 © David Muench/Corbis; p. 8 (left) © Ned Fiary/Lonely Planet Images; p. 8 (right) © Frans Lanting/Minden Pictures; p. 11 © Marc Muench/Corbis; p. 11 (inset) © Corbis; p. 12 © George D. Lepp/Corbis; p. 15 © Neil Rabinowitz/Corbis; p. 16 © Robert Essel NYC/Corbis; p. 20 (top left) © Michael and Patricia Fogden/Minden Pictures; p. 20 (top right) © Jose B. Ruiz/naturepl.com; p. 20 (bottom) © Jeff Foott/naturepl.com.

Library of Congress Cataloging-in-Publication Data

Nadeau, Isaac.
Caves / Isaac Nadeau.— 1st ed.
p. cm. — (Library of landforms)
Includes index.
ISBN 1-4042-3128-5 (lib. bdg.)
1. Caves—Juvenile literature. I. Title.

GB601.2.N34 2006
551.44'7-dc22

2005001555

Manufactured in the United States of America

Contents

Top: One of the largest chambers in the world is the Big Room in Carlsbad Caverns in New Mexico. The Big Room covers 8.2 acres (3.3 ha) and has a ceiling up to 255 feet (77 m) high. *Bottom:* Caves are often found in areas known as karst. Karst is made up mostly of limestone rock, which is a soft rock. Karst areas, such as the one shown here in Spain, are usually bumpy and full of openings called potholes and sinkholes.

What Is a Cave?

A cave is a hole in the earth that is big enough for a person to fit in. The smallest caves are the size of a small closet. The largest caves are hundreds of miles (km) long. Some caves are found in the sides of hills or mountains. There are also many caves in rock beneath the ocean. Caves are formed by natural methods, including the movement of water, wind, ice, and rock. Caves are found in many different types of rock, including **sedimentary rock**, such as limestone, and **volcanic** rock. Caves are also found in ice. Some of these caves include large rooms called chambers, connected by long, narrow tunnels.

In the United States, there are more than 17,000 known caves. There are many kinds of caves, but the largest and most common are found in areas with uncovered limestone, known as karst. In the United States, large karst areas can be found in Kentucky, Missouri, New Mexico, and Tennessee, among other states.

From the outside of a cave, it is hard to know how deep it is. This is because the sunlight does not reach all the way inside a cave. Caves are very different from the world above the ground. For example, in many caves the air does not get hotter or colder, even when such changes occur above ground.

Solution Caves

Solution caves are Earth's most common type of cave. They are usually found in limestone, which is considered a soft rock. Solution caves are formed when water **dissolves** rock. When rainwater falls and sinks into the ground, it gathers a gas called carbon dioxide from the plants in the ground. When water mixes with carbon dioxide, it forms an **acid** that helps the water dissolve rock. As the water sinks beneath the surface of the ground, it reaches a **layer** of rock called bedrock. When the water reaches bedrock, it flows along cracks and into any little holes it finds there. The water slowly dissolves the rock, and over thousands of years, the cracks and holes grow larger and larger. As holes and cracks grow larger, more water flows into them, causing the rock to dissolve even more. The more water that can flow into the holes, the more quickly they grow. Over time the water can create huge chambers and tunnels many miles (km) long. If the tunnel has an entrance from the surface, it is a cave. A solution cave with miles (km) of tunnels might have an entrance only a few feet (m) wide.

Lehman Caves can be found in the Great Basin National Park in Nevada. Lehman Caves first formed between three and five million years ago. It is considered a young cave. Even though it is called Lehman Caves, it is a single cave that is 2 miles (3.2 km) in length.

4.5 billion years ago: Earth forms. It is thought that Earth's first volcanic caves formed during this time.

3.5–3.8 billion years ago: Life on Earth begins.

About 600 million years ago: From this time to the present, limestone rock is formed from the bodies of small ocean animals. Many of Earth's caves are found in limestone rock.

10 million years ago: Mammoth Cave begins to form in the limestone area of Kentucky.

About 100,000 years ago: Earliest members of the human species live in caves in South Africa.

Today: All over the world, caves continue to be formed and destroyed.

A good way to remember the difference between a stalactite and a stalagmite is this. Stala<u>c</u>tites, with a *c*, such as the ones shown here on the left, drop down from the <u>c</u>eiling. Stala<u>g</u>mites, with a *g*, which are shown on the right, grow upward from the <u>g</u>round.

Formations in Solution Caves

Solution caves are decorated with many kinds of interesting rock formations, called speleothems. Stalactites and stalagmites are two of the most famous speleothems. Both are the result of water dripping from the ceilings of limestone caves. This water has dissolved limestone in it. As the water drips down, a tiny amount of the dissolved limestone can be left behind on the ceiling. As each drop leaves a little bit of rock behind, a little bump of limestone begins to form. As drops of water continue to drip from this bump, the bump continues to grow toward the floor, looking much like an icicle. These icicle-shaped formations are called stalactites. As the drips hit the floor below, more limestone is **deposited**, and sometimes a bump grows upward. This formation is called a stalagmite. Sometimes stalactites and stalagmites meet in the middle, forming what is called a column. Another speleothem, called a curtain, is formed when water slides down the curved ceiling and walls of a cave. As it flows down, it leaves behind deposits of stone, which build up over time. Some curtains are so thin that light can shine through them.

Volcanic Caves

Volcanic caves are caves found in volcanic rock. Volcanic rock is rock deposited during a volcanic **eruption**. Unlike solution caves, volcanic caves can be formed in a matter of just a few days. There are two different types of volcanic caves. **Lava** tubes are formed when the rock of a volcanic eruption is still in its liquid, or molten, state. As the lava flows down the sides of the volcano, the surface of the lava cools more quickly than the lava in the middle. As the lava on the surface cools, it hardens into rock, forming a tube of rock around the liquid lava inside it. This liquid lava continues to flow until it comes out the end of the tube, leaving behind an open cave called a lava tube.

Another type of lava cave is called a blister cave. Blister caves are shaped like giant bubbles. A lot of air can be trapped in lava as it erupts and flows down the side of a volcano. As the lava cools, some of this air rises through the lava like bubbles in water. In most cases these bubbles pop when they reach the surface. Sometimes the bubbles cool on the surface. The result is a thin-walled stone bubble, called a blister cave.

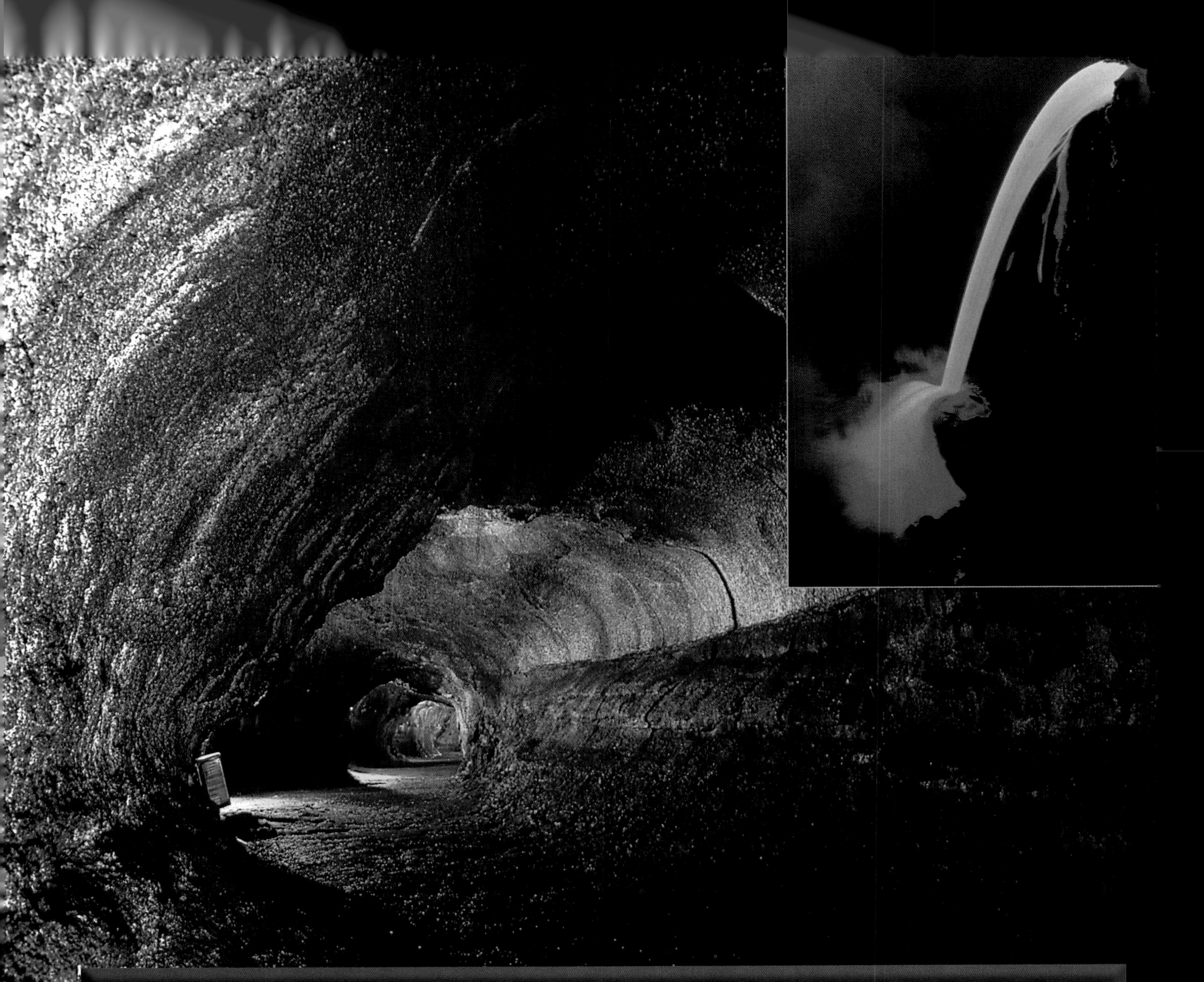

The Thurston Lava Tube is located on the Big Island in Hawaii. It was formed about 350 years ago and measures 400 feet (122 m) long. This is short compared to other lava tubes that can be miles (km) long. *Inset:* Here you can see lava flowing from a lava tube

Most glacier caves are found in Antarctic and Arctic regions. This glacier cave is found in Antarctica, near the South Pole. Glacier caves are formed when water in a glacier melts,

Glacier Caves

In very cold parts of Earth, such as in the mountains of Alaska or in Greenland, large parts of the land are covered with huge masses of ice, called glaciers. Glacier caves are caves found in glaciers. Each summer some of the glacier ice melts under the heat of the Sun. This melted water flows downhill over the surface of the glacier. When it reaches a crevasse, or a crack in the ice, it flows down inside the glacier. The water finally makes its way all the way to the bottom of the ice, where the glacier sits on top of the bedrock. There is no soil beneath a glacier, because it has all been **scraped** off by the ice. When the water reaches the bedrock, it continues to flow downhill along the surface of the bedrock. As it flows it slowly melts the ice of the glacier and forms a tunnel. Little by little the tunnel grows, until finally the water flows out the end, or snout, of the glacier. Throughout the summer glacier caves are filled with water that melted on the glacier's surface and flowed down into them. In the winter the surface of the glacier is frozen and no more water flows into the cave. Then it is possible for people to go inside the glacier cave and **explore**.

Caves Formed by Erosion

Sea caves and wind caves are formed by the forces of **erosion**. Sea caves are found in the rocky cliffs on coastlines. As waves crash against the rock of the cliffs, the rock erodes bit by bit. The places where the waves hit hardest erode more quickly than the places where they are more gentle. In weak places that are hit hardest by the waves, a hollow is carved, or cut out. It grows deeper with each crashing wave. Most sea caves are not very long. However, the longest known sea cave is Painted Cave on an island off the coast of California. The larger sea caves can be explored by boat.

Wind caves are often formed in desert cliffs. These cliffs are usually made of sandstone. They were formed long ago from bits of sand that piled up and pressed together to form rock. As wind blows across the desert, it picks up sand. When the wind blows against a cliff, the sand scrapes against the rock, carving out **shallow** holes. Over many years the holes become deeper, forming caves. Most wind caves are not very deep, however, since the deeper they get, the more **sheltered** they become.

The water level in sea caves rises and falls with the tides. At high tide many sea caves are completely filled with water. When the tide is low, people can often explore sea caves by boat

Compared with other landforms, such as mountains, most caves do not last very long. A mountain may stand for more than one hundred million years, but most caves do not last longer than a few million years. Some collapse when they are just a few thousand years old. Waves crashing against the coastline, as shown here, can help form and then destroy caves.

Caves over Time

Caves are always changing. In most cases the forces that create caves continue to shape them over time. Solution caves continue to grow as long as water flows into them from the ground above. Sea caves and wind caves grow as long as the erosion that shaped them continues.

Sometimes the same forces that create caves end up destroying them. An example of this is when a tunnel or chamber becomes so large that it caves in, or collapses, beneath the weight of the rock above it. This can cause the entrance to a tunnel to be blocked, or it can cause an entire cave to be filled in. This kind of collapse happens in all types of caves. Caves are destroyed in other ways, too. A glacier cave disappears when the ice of the glacier melts away. Lava tubes can become blocked when new lava flows into the tube and hardens.

Existing caves are always changing, but new caves are always being formed. As long as there is water, wind, and rock on Earth, there will be caves in formation.

Mammoth Cave

Mammoth Cave is a solution cave found near Cave City in western Kentucky. It is the longest known cave in the world. So far more than 350 miles (563 km) of tunnels and passages have been discovered in Mammoth Cave. Scientists believe that there are hundreds of miles (km) more that are not yet known. Some chambers in the cave are connected by tunnels so small a person can barely fit through them. If you were to explore Mammoth Cave, you might crawl through a narrow tunnel for hundreds of feet (m) before it opened up into a large chamber.

The cave was first discovered between 12,000 and 4,000 years ago by Native American explorers. They mined **gypsum** in the cave, which they used to make paint. They may also have used the cave as a shelter. In the 1800s, an African American man named Stephen Bishop explored and mapped many miles (km) of the cave. He was the first person to discover the underground rivers that flow in the cave.

Here you can see a stalactite formation in Mammoth Cave. Today almost 2 million people visit Mammoth Cave each year. They come to see the beautiful speleothems and the cave's many underground rivers. Some of these rivers, or waterways, are large enough for boats to float on.

Top Left: This tree frog lives in the entrance zone of a sandstone cave in western Australia. *Top Right:* Bats often use caves to hibernate, or sleep, during the winter. Some caves may have more than 100,000 hibernating bats hanging upside down from the ceiling during the winter. *Bottom:* This blind fish feels or hears its way through the waters in the deepest part of a cave in Mexico. This fish, called a blind tetra, is a troglobite.

Animal Life in Caves

Many different kinds of animals make their homes in caves. Caves can be separated into three zones, or areas. These are the entrance zone, the twilight zone, and the dark zone. The entrance zone includes the cave entrance and the areas of the cave that are touched by sunlight. Many animals live in or visit the entrance zone. They do not depend on the cave for **survival**. These animals are called trogloxenes. Trogloxenes in the entrance zone can include raccoons, people, and frogs. As you go deeper into a cave, there is less light. This is the twilight zone. Here there are some animals that spend their whole lives in the cave, and others that go in and out of the cave. These animals are called troglophiles. Earthworms are a good example of troglophiles. The deepest part of a cave, where it is completely dark all the time, is called the dark zone. Many of the animals found in the dark zone spend their whole lives there. These animals, called troglobites, have **adapted** to life without light. For example, there are fish and scorpions without eyes. Since there is no light in the cave, these animals have no need for eyes.

People and Caves

Scientists who study caves are called speleologists. Speleologists study the way water moves through caves and the rocks and natural elements in caves. People who explore caves for fun are called spelunkers. They wear hardhats with headlamps on them and often use ropes to make their way down steep tunnels. Some spelunkers even use boats to explore caves.

People first began using caves as shelter over 100,000 years ago. Much of what we know today about how people lived thousands of years ago comes from what has been found in caves. Human bones and the remains of fire pits can be found in caves, as can stone tools and bits of clothing. These objects help us know what kinds of things our **ancestors** ate, what they wore, and how they hunted and cooked their food. Cave paintings more than 30,000 years old have been found in caves in Africa. These paintings often include pictures of animals. By studying cave paintings, we can learn which kinds of animals lived near people thousands of years ago.

Today caves are a place for people to explore and study. Every year people discover new caves around the world. After more than 100,000 years of people visiting caves, there are still thousands of caves where people have never set foot. Learning about the science and beauty of caves on Earth is always an adventure.

Glossary

acid (A-sid) A liquid that breaks down matter faster than water does.

adapted (uh-DAPT-ed) Changed to fit new conditions.

ancestors (AN-ses-terz) Relatives who lived long ago.

deposited (dih-PAH-zuht-ed) Left behind.

dissolves (dih-ZOLVZ) Breaks down or fades away.

erosion (ih-ROH-zhun) The wearing away of land over time.

eruption (ih-RUP-shun) The explosion of gases, smoke, or lava from a volcano.

explore (ek-SPLOR) To go over carefully or examine.

gypsum (JIP-sum) A type of mineral found in the earth.

lava (LAH-vuh) A hot liquid made of melted rock that comes out of a volcano during an eruption.

layer (LAY-er) One thickness of something.

scraped (SKRAYPD) Rubbed or tore the surface of something.

sedimentary rock (seh-deh-MEN-teh-ree ROK) Layers of gravel, sand, silt, or mud that have been pressed together to form rock.

shallow (SHA-loh) Not deep.

sheltered (SHEL-terd) Guarded from weather or danger.

survival (sur-VY-val) Staying alive.

volcanic (vol-KA-nik) Having to do with a volcano.

Index

Web Sites

Due to the changing nature of Internet links, PowerKids Press has developed an online list of Web sites related to the subject of this book. This site is updated regularly. Please use this link to access the list:
www.powerkidslinks.com/liblan/caves